My First Adventures

# MY FIRST TRIP TO THE BEACH

By Katie Kawa

**Please visit our website, www.garethstevens.com. For a free color catalog of all our high-quality books, call toll free 1-800-542-2595 or fax 1-877-542-2596.**

**Library of Congress Cataloging-in-Publication Data**

Kawa, Katie.
My first trip to the beach / Katie Kawa.
p. cm. — (My first adventures)
Includes index.
ISBN 978-1-4339-7309-3 (pbk.)
ISBN 978-1-4339-7310-9 (6-pack)
ISBN 978-1-4339-7308-6 (library binding)
1. Beaches—Juvenile literature. I. Title.
GB453.K39 2012
551.45'7—dc23

2011043596

First Edition

Published in 2013 by
**Gareth Stevens Publishing**
111 East 14th Street, Suite 349
New York, NY 10003



Editor: Katie Kawa
Designer: Andrea Davison-Bartolotta

All Illustrations by Planman Technologies



Printed in the United States of America

CPSIA compliance information: Batch #CS12GS: For further information contact Gareth Stevens, New York, New York at 1-800-542-2595.

# Contents

I am going
to the beach!

I wear my new bathing suit. It is yellow.

The beach has lots of water. This is the ocean.

I walk on the sand.
The sun makes it hot!

I play in the sand.
I dig with a shovel.

I put the sand
in a bucket.

The water is warm.
It has a lot of waves.

I hold my mom's hand in the water. We jump in the waves.

I see lots of birds
at the beach.
They are called seagulls.

I had fun playing
at the beach!

# Words to Know

bucket

ocean

shovel



# Index